2 May 2016

ERRATUM

to

MCRP 6-11D

SUSTAINING THE TRANSFORMATION

1. Change all instances of MCRP 6-11D, *Sustaining the Transformation*, to MCTP 6-10A, *Sustaining the Transformation.*

2. Change PCN 144 000075 00 to PCN 147 000008 00.

3. File this transmittal sheet in the front of this publication.

PCN 147 000008 80

DEPARTMENT OF THE NAVY
Headquarters United States Marine Corps
Washington, D.C. 20350-3000

6 October 2014

FOREWORD

Since our 31st Commandant, General Charles C. Krulak, first published Marine Corps Reference Publication 6-11D, *Sustaining the Transformation*, in 1999, the Marine Corps has continued our proud tradition of making Marines, winning battles, and returning quality citizens to society. Like all previous generations, Marines today are equally as committed to our time-honored values of honor, courage, and commitment. Marines of the 21st century are among the finest we have ever forged; it is every Marine's duty to sustain that rich legacy. America trusts its Corps of Marines—we must always strive to preserve that trust.

The Marine transformation is forever ingrained in our DNA, from recruit training to Officer Candidates School and throughout the rest of our lives. The transformation to becoming a Marine is often the defining moment in a person's life. All Marines must possess a clear understanding that our Eagle, Globe and Anchor is much more than an emblem, rather that earning the title "Marine" carries a life-long responsibility to defend our nation, to care for our fellow Marine, and to keep our honor clean.

This publication is focused on one of our key leadership principles: *know your Marines and look out for their welfare*. It is a companion to Marine Corps Warfighting Publication 6-11, *Leading Marines*, and provides more detail to leaders about how to take care of their Marines through the five phases of the transformation process. It is a leadership tool that provides sound examples and methods for success. Sustaining a Marine through this process requires engaged leadership. I encourage you to use this reference as a starting point for discussions amongst peers, subordinates, and seniors. Your dialogue and application will improve the welfare of all Marines and our institution.

Sustaining our transformation is not a new concept; it has endured for over 230 years. The saying, "Once a Marine, Always a Marine" holds true because our transformations have been strengthened throughout our lives. We witness the health of our traditions and legacy in the proud smile of a new private graduating boot camp, or in the hearty handshake and greeting of two seasoned warriors. The Marine Corps family is alive and well, as is our rich heritage. I remain, *Semper Fidelis…*

Semper Fidelis,

JAMES F. AMOS
General, U.S. Marine Corps
Commandant of the Marine Corps

Publication Control Number: 144 000075 00

Table of Contents

Chapter 1. The Difference

Training Day 10 1-2

Training Day 15 1-2

Training Day 19 1-4

Training Day 27 1-6

Graduation. 1-6

3 Months Later 1-10

Chapter 2. Defining and Understanding Transformation

Transformation Impacts 2-3

Transformation Erosion 2-4

Marines Today. 2-6

Chapter 3. Phase I: Recruitment

Chapter 4. Phase II: Recruit Training

Chapter 5. Phase III: Cohesion

Understanding the Larger Mission 5-1

Forming *Esprit de Corps*. 5-2

Understanding Dimensions of Cohesion. 5-3

    Individual Morale 5-3

    Confidence in the Unit's Combat Capability 5-4

    Confidence in Unit Leaders 5-4

Horizontal Cohesion 5-5

Vertical Cohesion 5-6

Mutual Support of Horizontal and
  Vertical Cohesion 5-7

Chapter 6. Phase IV: Sustainment

Command Involvement 6-2

For the New Marine 6-2

For All Marines 6-6

The Marine Corps Family 6-8

Chapter 7. Phase V: Citizenship

Chapter 8. Critical Factors Affecting Sustainment

Obstacle Reduction 8-2

Strong Leadership 8-2

Be Technically and Tactically Proficient 8-4

Know Yourself and Seek Self-Improvement 8-5

Know Your Marines and Look
  Out for Their Welfare 8-6

Keep Your Marines Informed 8-6

Set the Example 8-7

Ensure Assigned Tasks are Understood,
  Supervised, and Accomplished 8-8

Train Your Marines as a Team 8-9

Make Sound and Timely Decisions 8-9

Develop a Sense of Responsibility Among
Your Subordinates 8-10

Employ Your Command in Accordance
with its Capabilities 8-11

Seek Responsibility and Take Responsibility
for Your Actions............................. 8-12

Interacting with Schools 8-13

Attending Unit-Level Corporal Courses 8-13

Encouraging Professional Military Education......... 8-13

Briefing New Joins 8-14

Maintaining Bachelor Enlisted Quarters............. 8-14

Providing Mentors............................. 8-14

Educating Leaders 8-15

Glossary

References

This Page Intentionally Left Blank

Chapter 1

The Difference

Private James Smith only saw his brother Private Tommy Smith twice while they were first in the Marine Corps—once in the chow hall during grass week at boot camp, and then again following graduation. However, both Marines did well during recruit training and were proud of themselves and of each other. This pride was evident in their eyes and in the eyes of their parents, Mr. and Mrs. Smith, at graduation. The Smiths felt both their boys looked and acted much older. After a short period of leave and some much-deserved rest, the Smith boys returned for training at the School of Infantry (SOI). During this training, they were in separate platoons. They saw one another occasionally, but only late in the day or on weekends.

TRAINING DAY 10

At a platoon meeting at the beginning of offense week, the squad leaders informed the privates of their future duty stations. James was going to 1st Battalion, X Marines. Since most of the platoon was going to the same battalion, the squad leader told them a little bit about their unit and where it was located. Private James Smith found out that 1/X was located at Camp X-Ray. He also was told that Marines from his new unit would link up with them later in the week. These Marines would observe training and do some initial counseling. That afternoon, the Smith boys crossed paths during offense "round robin" training. Tommy was told he would go to 2/X, which was located on the other coast. Both were disappointed that they were headed for opposite ends of the country, but they had known they would be separated sooner or later. They shook hands, wished each other luck on the 10K hike that night, then headed back to their squads.

TRAINING DAY 15

On the first day of military operations on urbanized terrain training, James was watching closely as his squad leader, Sergeant Brown, demonstrated window-entering techniques to the squad. When the squad broke into teams to practice the techniques demonstrated, James noticed several other Marines watching the platoon training. From a distance, they appeared to be a lieutenant, a staff sergeant, and several noncommissioned officers (NCOs).

They were talking to the squad leaders and appeared to be asking a lot of questions.

Later that night, during the security patrols, Private James Smith noticed one of these Marines, a corporal, travelling with his squad. He took notice of the NCO, but did not have time to pay much attention. Private Smith had been picked to be the patrol leader, and his focus was to move the patrol along its designated route. The patrol did well in the linear danger area they encountered. The other squad members obviously had listened to their squad instruction that week because they executed their short and long security patrols without errors. It was dark by the time the patrol was safely inside friendly lines and their squad leader critiqued Private Smith and his patrol.

Immediately following the brief, the squad broke up to refill canteens and eat chow. A figure approached Private James Smith in the dark. "Private Smith," the Marine said as he extended his hand. "My name is Corporal Wilson. I will be your squad leader in A Co, 1/X. Good job on your patrol tonight. Sergeant Brown has told me a lot of good things about you. You are going to make a good addition to the squad. Go grab some chow and water. We will have some time to talk later this week. Maybe I can answer any questions you have about the unit. Again, good job, and I will see you tomorrow." Private Smith replied, "Good evening, corporal," but nothing more. He was happy about what had just taken place. The next day, James ran into his brother on his way to the military operations on urbanized terrain fundamentals class. He told him what had happened the night before. He asked if his brother had seen or met his squad leader. Tommy looked disappointed and answered, "No, not yet."

TRAINING DAY 19

After live-fire rehearsals, James was called off to the side by Sergeant Brown. The sergeant told him to report to Corporal Wilson, who was sitting a hundred meters away on an MRE [meal, ready to eat] box looking through some folders. James hurried over and reported to the corporal. Corporal Wilson and Private James Smith spoke for some time. They discussed James' family, his training, his performance at SOI, and his physical fitness. Corporal Wilson seemed interested that James had a brother in the same class. Corporal Wilson leafed through the pages of the training folder that James had seen Sergeant Wilson use to track his test scores and training data.

After he closed the folder, Corporal Wilson began to tell James about 1/X. He told the young private that he was lucky to join a battalion with such a rich legacy. He told Private Smith of the battalion's accomplishments during World War I, of the numerous amphibious landings it had made in World War II, and how it held the line in Korea. In Vietnam, the battalion distinguished itself again in several battles and produced numerous Medal of Honor recipients)—three of which had been awarded to Marines in the same company that James would soon join. Corporal Wilson then explained, in detail, how the company was organized and gave an overview of the company's senior leadership. James was then told about the upcoming months of squad and platoon training, followed by two company field operations before the holidays. Corporal Wilson told James that he should be able to get home for the holidays, but that he would have to keep an eye on the duty roster. Following the holidays, the battalion would conduct a combined arms exercise in February and deploy that following summer to Okinawa, Japan. Private James Smith was extremely happy to hear about the overseas tour because one of the reasons he joined the Corps was to travel to foreign countries.

Once Corporal Wilson was finished, James had more information than he could remember. Corporal Wilson saw the overwhelmed look in his eyes and said, "Don't worry, there isn't a test on this tomorrow. I'll see you again at graduation and help you get settled into the company area. By the way, if your folks are coming to graduation, they are more than welcome to come over to the battalion area afterwards. We'll be passing out maps after the ceremony. They can head over to the battalion chow hall for noon chow where they can meet with the battalion commander and the sergeant major. Any questions for me?"

James could not think of any questions that the corporal hadn't already answered. He shook his head and said, "No, Corporal." Corporal Wilson said, "All right then, pay attention at your live-fire ambush patrols next week. One more thing, the battalion is sending over a truck the morning of graduation to help with your trash. So don't throw any of your gear away just because it won't fit in your sea bag. Put it on the truck and we'll get it over to the company barracks."

James went back to training feeling good and looking forward to the future. Although the training had been tough and almost 4 months long, he could see a light at the end of the tunnel and couldn't wait to join his new unit.

TRAINING DAY 27

At the water bull, halfway through the 20K hike, the Smith brothers met again. Private Tommy Smith still had little information on his new unit. The only thing he knew was that a gunnery sergeant from his unit would be out to talk to them.

On Friday, the gunnery sergeant spoke to Tommy and about 25 other Marines *en masse* who were headed to 2/X. He went over the unit's history and schedule. When he was done, Tommy was dismissed, and the gunny then spoke to seven of the Marines who were headed to Company F. Tommy was slated for Company G.

GRADUATION

The day that both brothers had been looking forward to for so long had finally arrived: graduation day. Their parents had travelled a long way to see their sons' graduation and to wish them well as they moved on to yet another adventure.

The graduation ceremony was full of music and Marine Corps customs. Officers and staff NCOs were dressed in their Service C uniforms. James noticed the same officer and staff NCOs he had seen in the field, Corporal Wilson, and several other NCOs were sitting up front. When the guest of honor was introduced, James swore he had heard the lieutenant colonel's name before. He then heard the announcer say that the lieutenant colonel was the commanding officer of 1/X, his unit. James thought the lieutenant colonel had a hard, tough facial expression and graying hair that caused him to look older than his dad. But the lieutenant colonel looked like he was in good shape. His commanding officer praised the Marines for the hard work that had brought them this far, and he congratulated them on their accomplishments. He welcomed those new Marines who were coming to his battalion, and he went over some of the battles and victories Corporal Wilson had discussed with James during their meeting. After the speech, the commanding officer invited the families to lunch in the battalion area and told them the graduates would join their families in the afternoon.

Upon conclusion of the ceremony, the Smith brothers were exhilarated and quickly made their way over to their folks. After some time was spent slapping backs and posing for photos, the boys knew they needed to grab their bags and move out. Mr. Smith said he had a map to James' battalion area and would meet him after lunch. Mr. Smith turned to Tommy and asked where he would be. With a look of disappointment, Tommy answered that he was getting on a bus in a half an hour that would take him down to the airport for a flight to his new battalion located on the other coast. The Smith brothers realized this would be the last time they would see each other for some time and said their goodbyes.

James went to his barracks. His roommates told him the truck outside was from 1/X and could be used for their gear. James threw his bags on the truck and made his way to the bus where Corporal Wilson was standing. Corporal Wilson saw Private James Smith walking up. "Hey, Smith, congratulations," Wilson said, "Do you have all of your gear on that vehicle?" "Yes, Corporal," James answered. "OK, climb aboard," Corporal Wilson said.

As James sat down, he saw his brother getting on another bus on the other side of the parking lot. Tommy had his head down as he climbed aboard the bus bound for the airport.

It was a short ride for James to his battalion area. Corporal Wilson pointed out several ranges, training areas, and hike routes along the way. When they arrived at the barracks, several Marines were waiting for the buses. Corporal Wilson took the six Marines assigned to his squad and led them to their rooms after they picked up their gear from the truck. James' name and the names of his roommates were already on the door when they arrived. Clean linen was folded neatly on their racks along with a folder labeled "Information Packet 1/X."

Corporal Wilson said they had an hour to unpack and make their racks. In an hour, he would meet them out front and take them to the chow hall to see their families. After the corporal left, James opened the information packet. It contained a base map; bus routes and schedules; and information about the gym, banks, barber shops, the chow hall, and other facilities.

An hour later, Corporal Wilson appeared outside. He formed up his squad and marched them about a quarter mile to a large parade deck. On it were several large tents with tables containing food and refreshments. Surrounding the parade deck were various tactical vehicles mounted with weapon systems. Several Marines were dressed in utilities with full field gear and they had weapons that included the M-16, an 81-mm mortar, and a .50 caliber machine gun. Corporal Wilson pointed out the battalion headquarters and chow hall. He said their families would be finished with chow shortly and would meet with them on the parade deck. He told the Marines they had 2 hours before evening formation to look at the static displays, tour the battalion area, and show their families their barracks. Once Corporal Wilson dismissed the Marines, James found his parents at one of the demonstration sites. His father told him that the battalion commander spoke again at lunch, along with the sergeant major and the chaplain. Private James Smith spent the next 2 hours with his parents looking at the weapon demonstrations and touring the headquarters and barracks. That evening, James said goodbye to his parents and he made his way to evening formation.

3 MONTHS LATER

Private First Class James Smith was promoted this morning. His entire squad congratulated him immediately following the company formation. During that same formation, Corporal Wilson received his sergeant chevrons. Private First Class Smith was excited and could not wait to tell his brother Tommy and his parents about his accomplishment.

James hadn't heard from his brother for some time. Tommy used to call often, but he never sounded happy. Since Tommy never had much to talk about, James would fill the conversation with stories about Sergeant Wilson, his platoon, and the Marines in his squad. When James asked Tommy what was wrong, Tommy would only reply that James was lucky.

James called to tell his brother about his promotion. Tommy didn't react well to the news. James asked Tommy when he'd be promoted to private first class. Tommy said he didn't know, that he had gotten into some trouble about a month ago and he wasn't sure what would happen. James tried, but he couldn't get Tommy to tell him what happened. Tommy would only say that he had made some bad choices since joining his unit, that he had made friends with some of the wrong people, and he had paid the price. Tommy told James that he was upset with what he had done and that he wanted to turn himself around and be the type of Marine they always dreamed of becoming. When James hung up, he felt bad about their conversation and he was concerned for his brother.

James then called his folks to pass on his good news. They were excited to hear about James' promotion. They told James about the letter they had received from his platoon commander stating what a fine job he was doing and his upcoming promotion. His parents said it had been some time since Tommy had called home, and they were starting to worry. They asked if James had heard from Tommy lately. James assured them that Tommy was okay and that they had recently spoken.

James hung up from talking to his parents. He felt a little down because he was worried about his brother, and he wondered what could have happened to him and how he could have gotten into trouble. Both of them had wanted to be Marines since they were kids and couldn't wait for the day they could wear the Eagle, Globe, and Anchor. Now, only 3 months after arriving at his unit, Tommy wasn't hacking it. Although Private First Class Smith was bothered by the news of his brother, he had little time to waste. Sergeant Wilson had assigned him to teach a class on crossing linear danger areas during the squad patrolling exercise the next day. James was excited about the opportunity to teach his fellow squad members and proud that Sergeant Wilson had picked him to do it. He turned to his desk and the stack of class handouts and field manuals. He would spend the evening reviewing his handouts and creating a class outline that would make Sergeant Wilson proud.

This Page Intentionally Left Blank

Chapter 2

Defining
and Understanding
Transformation

Transformation, as Marines define it, is the ongoing, dynamic process that begins with the prospective Marine's first contact with the Marine recruiter and continues through the Marine's entire life. It has five distinct phases: recruitment, recruit training, cohesion, sustainment, and citizenship. These phases—unique, but linked—are detailed in the chapters of this publication and ensure transformation remains a lasting and positive experience throughout every Marine's life.

The transformation process—going from civilian life to a Marine and one day returning to civilian life changed as a Marine—is difficult to codify because it takes on a personal and different perspective for each Marine. There are different reasons young Americans raise their right hand to take the oath and enter our Marine Corps: many are motivated by a deep sense of patriotism, some want challenging adventures in an elite organization, and some are looking to better themselves as a foundation for a later and much different career.

Reasons for leaving our active ranks are just as diverse and personal. Most Marines leave by their own choice, having served honorably at the end of their first term. Some must leave our active ranks for various other reasons, while others remain in our ranks for decades to become the senior leadership, guiding and developing younger generations of Marines. These reasons for joining, staying, or leaving the ranks can often fluctuate or change as a Marine experiences different units, challenges, responsibilities, and personal life-changing events. What we must not allow in any phase of transformation are Marines who feel they are disillusioned, disheartened, or forgotten. "Proud to claim the title" and "to keep our honor clean" must resonate powerfully in all who have worn the Eagle, Globe, and Anchor.

Despite the personal differences for joining or staying in the Marine Corps, there is a common spirit that bonds each Marine together in deep emotional fellowship—from the time we earn the title until we are buried—regardless of what Marine generation we lived in and how long we served. That earned fellowship is at the heart of why we say, "Once a Marine, always a Marine." Our mutual admiration and respect for each other is the manifestation of shared, positive transformation processes. There is no such thing as an "ex-Marine" or "former Marine." We are either active duty, reserve, or veteran Marines.

TRANSFORMATION IMPACTS

It is important to recognize that the success of each individual's transformation phase positively impacts not only the Marine but also his fellow Marines. Successful transformation reaches outside the ranks to the Marine's family and into the community where the Marine lives. Marines thriving in each positive phase of transformation help enable success in their personal lives. It is this time-tested success of transformation that is the essence of what Marines are, what we stand for, and how we interact in our professional and personal lives. Transformation is the collective soul of our Marine family passing from generation to generation that binds us together and renews us. The legacy of the Marine Corps is ultimately judged by the entire transformation experience of all Marines and that legacy influences the next generation of Americans that will join our ranks to start the cycle all over again.

The bond of Marine fellowship is tested continually within each Marine and tested in each phase of the transformation. The quality of those phases is a pact between the individual Marine and the Marine Corps. The strength of transformation from phase to phase in a Marine's experience can be greatly enhanced by his fellow Marines or become a negative experience that can leave that Marine and/or his family feeling alienated or deeply disillusioned, breaking the transformation bond for the rest of that Marine's life. All Marines must be proactive in order to keep each other's transformation strong. At the first sign of a strained transformation in a Marine, we owe it to each other to take immediate, corrective, and tangible actions to help that Marine. Such action is the hallmark of involved and engaged leadership and such leadership nurtures and develops the total Marine.

TRANSFORMATION EROSION

There are specific areas that can quickly erode a positive transformation in any Marine wearing the uniform. Those areas, if undiscovered or left alone, will quickly affect the morale and effectiveness of the Marine and his unit, and possibly leave a stain on our respected heritage. Those areas are—

- Lack of good order and discipline in garrison or when deployed.
- Lack of physical fitness.
- Poor personal appearance.
- Lack of accountability.

Other areas that often result in very serious consequences are fraternization, sexual harassment, sexual assault, hazing, and substance abuse.

Most problems Marines have can be traced back to a breakdown of leadership and discipline. All Marines have a responsibility to immediately take action to correct these transformation problems, if detected. They must not be tolerated in any Marine or that Marine's leadership. Although these transformation problems are not addressed in detail in this publication, this publication is a foundational starting point for Marines or their leaders to get help for Marines faltering in transformation. There are resources at the unit, base, or station as well as on the Internet that are available for all Marines.

An individual Marine's actions can have a profoundly positive or negative impact on our legacy due to the "connected" world we live in. Instant media and the Internet are not a good place for our young Marines if they do not have a good foundation. A Marine can seriously embarrass, endanger, or cause difficulties for both himself and the Marine Corps; for example, careless posts on blogs, Web pages, or in the social media can bring shame on the Marine or to our Corps. Though sometimes not intended to cause the Marine Corps embarrassment or endanger Marines, thought-less comments or photos have given away troop positions, future operations, and sensitive information, or brought discredit on our Corps. Some of these social media posts have made Marines and the Marine Corps appear reckless and dishonorable, or, at the least, call into question our strong core values of honor, courage, and commitment. Marines must self-censor their online presence, and they are expected to conduct themselves as they would in person and in uniform.

We must continue to do everything possible to retain the respect and admiration for the United States Marine Corps by our fellow Americans and by our Marine family. The following chapters identify engaged leadership actions to help ensure quality trans-formation in all five phases for Marines. It is not meant to be all inclusive, but is a foundational reference for further discussions and detailed actions. Our Marines have always been innovative no matter the challenge, that is no different when taking care of our fellow Marines. These tangible actions and resources must be coupled with the intangible aspects of Marine Corps leadership/ mentorship to be successful. Marines must rely on each other, their oath, their experience, and their own conscience to help guide their actions about what is right for the Corps.

MARINES TODAY

Honor, courage, and commitment are values that *All Marines* must inculcate and demonstrate in their every action. We must ensure that every Marine has been armed with the courage to confront the confusing, chaotic battlespace of the 21st century. We must continue to "issue" every Marine a true compass of personal honor and the commitment to forego interests of self for the interests of their comrades, their Corps, and their Country.

Marines today are challenged to operate in the complex environments of peace, crisis, and war, all within a very short period of time or simultaneously during the same operation. This promises to be commonplace in future conflicts. Today's battlefield is increasingly technologically super-charged, often ill-defined, and compounded by the confusion of modern warfare characteristics. In addition, Marines have taken on greater responsibilities in humanitarian and security cooperation missions that also may be ill-defined and chaotic. Today's operations require rapid, more complex decisionmaking at lower levels and place great stress on the individual Marine to be decisive, adaptable, innovative, and resilient.

Because of the complex operating environments, we need to prepare our young Marines for future battles. Dealing with decentralized operations, advanced technology, increased weapons lethality, asymmetric threats, mixed combatants and noncombatants, and urban combat is the operating norm so far in the 21st century. To succeed in a changing operating environment, our Marines must be good decisionmakers, highly trained, and self-confident. They must have absolute faith in the members of their unit, and their unit must have faith in them. We must ensure that our newest Marines fully understand and appreciate what the Marine Corps represents, and, by becoming members of the world's fighting elite, they uphold the sacred trust we have with our great Nation and with each other.

Transformation is an ongoing, dynamic process that begins with an individual's first contact with a Marine recruiter and continues throughout a Marine's life. We must ensure that, as Marines grow and life changes their circumstances, the Marine Corps will help them find the tools they need to transition back to a productive and faithful citizen. For about 70 percent of our Corps, the five phases of transformation will be completed within a one-term enlistment of four years. We are a young Service and our noncommissioned officers play a huge role in guiding our youngest Marines—now more than ever.

The foundation for the transformation was laid in recruiters' offices all across America, in the squad bays of our recruit depots, and at our formal entry-level schools. It is only a foundation, one which the Marine Corps and each individual Marine must make an effort to build. There are four pillars built on this foundation upon which the transformation rests: education, NCO development, ownership and acceptance, and establishment and maintenance of standards.

We must continue to educate the Corps about the transformation process. Leaders of Marines must understand the process that developed their Marines. Armed with this knowledge, these leaders will understand the benefits gained through the transformation process and the true capabilities of their Marines.

Significant to sustaining the transformation is selecting the best Marines to be NCOs, and continuing to train them to sustain the values and warfighting ethos of our Corps. Marine Corps NCOs are the primary leaders of our first term Marines and must emulate the high standards of what small unit effective leadership requires. We put a great deal of trust in our NCOs and they must sometimes operate without direct higher guidance. They greatly influence a young Marine who may be deciding whether to stay in our Marine Corps for a career. More importantly the first term Marine looks to the NCO ranks as mentors.

Regardless of rank, every NCO, staff NCO, and officer must personally accept the mission to sustain the transformation of the Marines in their charge and those Marines they encounter throughout their daily routine. Every Marine must take personal ownership of the transformation and commit to it as a way of life.

Chapter 3

Phase I: Recruitment

The first phase of the transformation process begins with our potential Marines meeting with our recruiters. Recruiters carefully screen young people who come to our door seeking admittance. Those with solid character, good moral standards, and personal values are those we embrace and validate, reinforcing the values they hold. Those with undamaged characters, but who are among our society's many "empty vessels," we fill with the ideals and values they so desperately need and seek. We evaluate each candidate based on the whole person and decide on acceptance or rejection through an analysis of risk versus potential.

Meanwhile, candidates are evaluating the Marine Corps based on what they perceive the Marine Corps to be and their personal experiences. Hence, the individual and the Marine Corps evaluate each other at this first phase of transformation to ensure each understands what is to be gained and expected, what they are joining, and what they are expected to become. The Marine recruiter is a mentor and launches a recruit's transformation. The potential recruits, also known as poolees, are better prepared when they reach recruit training because they receive their first introduction to our core values, enhanced physical conditioning, knowledge of our history and traditions, and study guides that facilitate their transition from phase I to phase II of transformation.

A well led pool is a healthy pool. A healthy pool can sustain itself and contribute to success in the same manner as a Marine Corps unit. It increases the success of the recruits at the depots, which, in turn, facilitates and enhances the quality of Marines arriving at the operating forces.

Entry into the pool is an invitation to a seat at the table for the poolee. The way we treat the poolee is the first step of the transformation. It is designed to put individuals who have enlisted in the Marine Corps in a structured environment that provides monthly, weekly and, sometimes, daily contact with their recruiters and fellow poolees in preparation for basic training. It builds camaraderie.

A well-led pool requires weekly and monthly activities designed to ensure success—mental, physical, spiritual, and social—at basic training. The activities may include weekly physical training, individual mentorship sessions with recruiters, close-order drill practice, and Marine Corps knowledge development. All of these activities are used to increase the preparedness and confidence of the poolees who reach recruit training. A poolee from a healthy pool that focuses on preparation for success at recruit training feels taken care of by the Marine recruiter and thrives on a bond formed with fellow poolees; fellow future Marines.

The impact recruiters can have on a poolee is lasting. For example, upon first entering the pool, a poolee finds himself struggling to complete the initial strength test: a modified PFT that gauges if a poolee is physically fit enough to attend basic training. After spending 5 months of continually working with his recruiter to strengthen his upper body and increase his stamina, the poolee was ready. He felt like he was a lot more prepared for recruit training and felt supported by that recruiter. The poolee's confidence had sky-rocketed and he felt that he could now do things he never knew he could do. The poolee discovered he could push himself a lot further than he thought possible with the help of his recruiter. That poolee will never forget his recruiter because of the encouragement and personal time the recruiter spent helping him strengthen not only his body but also his confidence.

The relationship between a recruiter and the pool does not end when the poolee ships to recruit training. A recruiter who believes in sustaining the transformation and making the Marine Corps stronger through each and every Marine will communicate with the poolees throughout their recruit training, congratulating successes and encouraging them through their setbacks via mail and phone calls.

A recruiter has accomplished the mission when, years later, Marines still remember, and even reach out to, the recruiter when facing the end of the first term with re-enlistment on the horizon.

This Page Intentionally Left Blank

Chapter 4

Phase II: Recruit Training

The second phase of transformation takes place during recruit training. During this phase, we prepare all Marines—male and female, those destined for combat arms, and those destined for combat service or combat service support—to fight on the nonlinear, chaotic battlefield. During the second phase, the drill instructor becomes the central person to transform the young recruit's life. The drill instructor is the backbone of the recruit training process, a role model as recruits are immersed into Marine Corps culture and values. The drill instructor embodies the epitome of a Marine's physical strength, mental strength, spirit, and character. It is the drill instructor's responsibility to be the pivotal role model, leader, and mentor of these young men and women; to show them how to function as a team; and to teach them to persevere.

Recruit training is a 12-week transformation—a rite of passage—culminating in the Crucible, a 54-hour continuous test of intense, physically-demanding training under conditions of sleep and food deprivation. During recruit training, the Crucible is the defining moment for the recruit. It will not be the hardest challenge

Marines face in their entire lives, but, for most, the Crucible will be the first time they reach the limits of their mental, physical, and emotional endurance.

They will know that they are capable of much more than they previously believed. They will know that they can exceed their physical and mental limitations through teamwork, perseverance, and courage. Once experienced, the Crucible becomes a personal touchstone that demonstrates the limitless nature of what they can achieve individually and, more importantly, what they can accomplish when they are part of the Marine Corps team.

The drill instructor's job is not over when his recruits complete the Crucible. Week 12 is known as transition week, when recruits have the opportunity and the responsibility to increase their knowledge and confidence. Much of the transformation process is built during recruit training, but it is only the beginning of a lifetime of being a Marine.

Chapter 5

Phase III: Cohesion

The third phase of the transformation process is cohesion, strengthening what was born during recruit training. It is cohesion that binds Marines together. In this phase, Marine leaders in the operating forces make their first impressions on our newly joined Marines. Setting a tone that ensures our new Marines know they are genuinely welcomed, properly cared for and are quickly indoctrinated into their unit is what this phase is all about. We define cohesion as the intense bonding of Marines that is strengthened over time, resulting in absolute trust, subordination of self, understanding of the collective actions of the unit, and appreciation for the importance of teamwork.

UNDERSTANDING THE LARGER MISSION

Unit cohesion increases fighting power, provides positive peer pressure, and reinforces our core values as the team's collective sense of honor becomes dominant over self-interest. Marines train together, garrison together, deploy together, and fight together. A leader who receives a new Marine must ensure that he or she is properly sponsored and coalesced into the unit. Cohesion cannot simply be among peers: of equal importance is the manner in which individual Marines identify with their units. All leaders must make unit cohesion one of their highest priorities and principal objectives.

Marine leadership, at all levels, must strive to ensure that all their Marines know how they fit in to the "big picture" of the mission of the unit. In general, people seem to work with greater enthusiasm when they understand why their work is important and how it accomplishes the end goals.

FORMING *ESPRIT DE CORPS*

The more we reinforce the cohesion of our units and encourage an *esprit de corps*—a common spirit of comradeship, enthusiasm, and devotion—the stronger our units will be and the easier it will be to reinforce Marine Corps individual core values through positive peer pressure, mentoring, and leadership.

A good example of this *esprit de corps* is when an individual Marine risks his life to aid a fellow Marine or to accomplish the mission at hand. An example from our past includes the cohesion that bound Presley O'Bannon and his few Marines together during their march across 600 miles of scorching desert to stand triumphant at the shores of Derna. Another example is that of a trapped Marine division that bravely fought its way across frozen Korea, through six communist divisions, to the sea. That legacy is further evidenced in the more recent conflicts of the past decade. Battles in places, such as Fallujah, Ramadi, and Helmand Province are replete with extraordinary heroism of units under extreme and challenging combat. Cohesion provides Marines with supportive relationships that buffer stress and increase their ability to accomplish the mission or task. Strong unit cohesion results in increased combat power and the achievement of greater successes.

UNDERSTANDING DIMENSIONS OF COHESION

There are five dimensions of cohesion—individual morale, confidence in the unit's combat capability, confidence in unit leaders, horizontal cohesion, and vertical cohesion. In combination, these dimensions dramatically affect the capabilities of a unit.

Individual Morale

Leaders must know their Marines and look out for their welfare. As Sir William Slim reinforced to the officers of the 10th Indian Infantry Division, "individual morale as a foundation under training and discipline, will bring victory." (*Dictionary of Military and Naval Quotations*) Marine leaders who understand these simple words are more likely to keep morale high among individual Marines. A high state of individual morale, in turn, enhances unit cohesion and combat effectiveness.

Confidence in the Unit's Combat Capability

Marines' confidence in their unit's combat capability is gained through unit training. The longer Marines serve and train together in a unit, the more effective they become and the more confident they are in their unit's capabilities. They know what their unit can do because they have worked together before. Keeping Marines together through unit cohesion is a combat multiplier. Rarely are battles lost by those who maintain confidence in their unit and in their fellow Marines. Success in battle can be directly attributed to a unit's overall confidence in its level of performance. Of course, the opposite also holds true: lack of cohesion, lack of confidence, and poor performance preordain a unit's failure. As Brigadier General J. B. Hittle asserted, "If the history of military organizations proves anything, it is that those units that are told they are second-class will almost inevitably prove that they are second-class." (*Dictionary of Military and Naval Quotations*)

Confidence in Unit Leaders

Confidence in unit leaders' abilities is earned as Marines spend time in the company of their seniors and learn to trust them. Leaders must earn the respect of their Marines, and doing so takes time. As Marines develop confidence in their units' ability to accomplish their assigned missions based on their prior achievements, they also develop confidence in their leaders as they work and train together. Major General John A. Lejeune believed that—

The relation between officers and men should in no sense be that of superior and inferior nor that of master and servant, but rather that of teacher and scholar. In fact, it should partake of the nature of the relation between father and son, to the extent that officers, especially commanding officers, are responsible for the physical, mental, and moral welfare, as well as the discipline and military training of the young men under their command. (*Dictionary of Military and Naval Quotations*)

Horizontal Cohesion

Horizontal cohesion is as important on the asymmetrical battlefield of today as it was in the island-hopping campaigns of World War II. Horizontal cohesion, also known as peer bonding, takes place among peers. It is the building of a sense of trust and familiarity among individuals of the same rank or position. Sense of mission, teamwork, personnel stability, technical and tactical proficiency, trust, respect, and friendship are some elements that contribute to peer bonding.

An example of horizontal cohesion is the relationship between members of a fire team. Over time, each member develops a sense of trust in the other. This trust is born of several elements. The first is a common sense of mission, the act of placing personal goals aside to pursue the goals of the entire team. Other elements include teamwork and personnel stability. Teamwork is the result of mutual support provided by each member of the team.

Teamwork is further enhanced by personnel stability, which promotes familiar and effective working relationships. Perhaps most important is the development of tactical and technical proficiency that continues to support and reinforce the trust and respect among the team members. When Marines are thrust deep into the chaotic battlespace, often operating in small teams, their will to fight and ultimately succeed will hinge upon their ability to fight as an effective, cohesive team.

Vertical Cohesion

Vertical cohesion is not new to our Corps; this dimension of cohesion involves the relationship between subordinate and senior. Vertical cohesion is what draws peer groups into a cohesive unit, such as a battalion or squadron. It is, in part, the building of a mutual sense of trust and respect among individuals of different rank or position. Additionally, vertical cohesion is the sense of belonging that the squad or section maintains relative to its role in the battalion or squadron.

An example of vertical cohesion is when many squads and sections come together to form a cohesive company. Each of these subordinate units plays a different role in the company; however, vertical cohesion draws them together in purpose and mutual support. This sense of unity has several elements:

- A common sense of unit pride and history—not only in themselves as a unit, but also in those who have gone before them. The organizational memory of their past achievements drives the unit to still greater heights.

- Quality of leadership and the command climate in the unit. Vertical cohesion is stronger in units with effective, well-trained leaders. Leaders who show concern for their Marines and lead by example will earn the trust and respect of their subordinates.
- Shared discomfort and danger, which can occur during shared training.

Mutual Support of Horizontal and Vertical Cohesion

Since the birth of our Corps, Marine units have experienced horizontal and vertical cohesion to varying degrees and with varying success. However, it is vital that these two qualities be developed in combination with each other. Just as the strength of combined arms comes from the combined effects of two or more different arms that mutually support one another, the strength of horizontal and vertical cohesion comes from the combined effects and mutual support they provide each other.

Blending vertical and horizontal cohesion ensures a strong, universal sense of bonding and teamwork among various types of units. If vertical and horizontal cohesion are mutually supported, all these units will be composed of Marines who trust and respect each other. Each type of bond reinforces the other. A cohesive battalion that comprises cohesive companies that place the goals and interests of the battalion or company above those of their squad and/or section is an example of both vertical and horizontal cohesion. A unit capable of combining vertical and horizontal cohesion is far stronger than a unit that is strong only in one.

This Page Intentionally Left Blank

Chapter 6

Phase IV: Sustainment

The fourth phase of transformation is sustainment. Sustainment is continuous and is part of everything the Marine Corps does. Our professional military education schools are designed to educate our leaders—our officers, staff NCOs, and NCOs—in "whole Marine" character development. Leaders in the operating forces and in the supporting establishment accomplish their missions in ways that support and reinforce our core values and foster team building. Leaders will manifest our core values and mentor their subordinates, living the Marine Corps ethos through a shared responsibility for all Marines that lasts even after a Marine is no longer in uniform. The principles of inclusion, recognition, and family and community outreach apply to sustaining a Marine's transformation regardless of rank.

COMMAND INVOLVEMENT

There are many ways to sustain the transformation, but almost all involve command participation and interest in each Marine. The following subparagraphs offer ideas that have been used successfully within Marine Corps units.

For the New Marine

The sooner a unit can establish contact with incoming Marines the better.

At School. If possible, contact should be made at the SOI or the military occupational specialty (MOS) school. Units can include their Marines who are undergoing MOS school training in the following ways:

- *Interaction with MOS schools.* The receiving unit coordinates with the MOS school to send unit leaders to observe training, meet their new Marines, and provide unit information.

- *Command liaison.* If squad leader interaction is not possible for every unit, a command liaison will suffice. The liaison provides new Marines reporting to the unit with such information as the schedules and billeting information.

- *Unit history brief.* The unit history brief provides new Marines with vertical cohesion at the earliest stages. A sense of pride develops as these young Marines realize they are becoming part of a rich, valiant legacy. Unit commanders should take the time to recognize the anniversary of a significant event or battle in which the unit participated. A good unit history program, complete with recognition of battles won, accolades earned, and sacrifices made, can also increase vertical cohesion.

- *Schedule/deployment brief.* Many young Marines look forward to their deployment overseas. Unit schedules and deployment briefs are motivators for Marines and also assist them in mentally preparing for the future.

At Graduation/Unit Reception. Command involvement at the SOI or MOS school graduation can be very beneficial. If unit leaders participate in the ceremony, graduation becomes an opportunity to initiate vertical cohesion with their Marines. The following subparagraphs discuss other ways to use graduation and receptions to foster vertical cohesion. These should be executed expeditiously to ensure a seamless transition into the unit.

Unit Reception. If possible, after graduation, the new Marines, and their gear, should be transported to the unit area by their new unit. The new squad leaders who have participated in training with these Marines can guide them through the check-in process. Upon arrival at their new unit, the new Marines should be welcomed to the unit by the Commanding Officer or Sergeant Major. The new Marines and their families can participate in a command-sponsored meal at the chow hall. Families may tour the unit area and billeting. Some units conduct static displays of various weapon systems and equipment to educate the families. These new Marines will also be guided through their administrative and logistical in-processing in an efficient manner. This is a team effort that will require support from the entire unit.

Unit Information Packet. Unit information packets ease the transition of Marines into their new units and provide answers to the many questions that result from relocating to a new environment. The packet also should include information on the unit's history, traditions, and future deployment schedules.

Hometown News Releases. Hometown news releases assist in developing vertical cohesion. At a minimum, a news release should address the Marine by name and rank, identify his unit, and identify his accomplishments. For example, "PFC John Smith, having completed 4 months of intense training, has earned the privilege of joining 1/X, a highly decorated unit with a proud Marine Corps history. PFC John Smith will be a welcome addition to this proud unit."

Command Letter to Families. Many families will not be able to attend the graduation or unit reception. A command letter may be the best way to ensure that all families are contacted. This letter should welcome the extended family and ensure them that their family member will be cared for.

Assignment of a Mentor or a Sponsor. Upon arrival to a unit, Marines must be assigned a mentor who can assimilate them into the unit. The mentor can enhance both vertical and horizontal cohesion.

In-Briefs. Many units will receive small numbers of individuals and cannot conduct a detailed unit reception. In this case, units must establish in-briefs that are conducted at regular periods in order to reach new Marines shortly after entering the unit.

One-Month Recognition. One month after Marines arrive at a unit, the command should recognize their contribution to the unit. This recognition increases Marines' vertical cohesion to the unit and command. Units can do this by giving awards to the most deserving Marines, such as a meritorious mast or a nonstandard, unit-specific award. Examples of nonstandard unit awards might be unit coins or T-shirts. This is another opportunity to send a command letter home in order to reinforce the positive behavior exhibited by the new Marines during that crucial first month period.

Six-Month Recognition. The 6-month recognition is similar to 1-month recognition; however, this is the last time that new Marines will be identified or addressed individually in this manner. After 6 months, they should have participated in exercises and operations and shared hardships and other experiences with the unit. A 6-month recognition can include events used during the 1-month recognition. It can also include, but is not limited to, the following recognitions:

- *Meritorious promotion.* A meritorious promotion is an opportunity to recognize those new Marines who have excelled since their arrival.

- *Command logbook.* During this ceremony, a battalion logbook should be on hand. This book would have inscribed within it the names of all the Marines who have belonged to that battalion. Each new Marine having honorably served with the unit for at least 6-months would sign the book.

As soon as a Marine gets promoted for the first time, he is a leader. That fact should be understood by the Marine and treated as such by his own leadership.

For All Marines

While special attention should be paid to new Marines to make them feel welcome and accustomed to their new lives as Marines, it is important to remember that most Marines, regardless of rank, will be "new" to a place every few years as they move through permanent change of station cycles. There should be an effort to welcome all Marines joining the new place. This is also the responsibility of the commander to foster an environment of inclusion and recognition.

Unit Events. The following are examples of events units can do regularly to maintain positive morale and build cohesion:

- *Field meets.* Unit sporting events are time-proven methods that develop both horizontal and vertical cohesion.

- *Unit special orders.* Mentioning a unique and important contribution of a Marine or a unit of Marines at a unit formation provides visible recognition of the accomplishment on behalf of a unit's mission. Unit special orders are read at formations by the unit's commanding officer, executive officer, adjutant, or sergeant major.

- *Marine Corps Birthday Ball.* Because the Marine Corps Birthday Ball is rich with tradition, it is a perfect opportunity to build vertical cohesion. Commanders must ensure that the birthday ball is an affordable and enjoyable event filled with honor, history, and tradition.

• *Marine's birthday.* Recognition of a Marine's birthday is another way to reinforce vertical cohesion at the small-unit level. The acknowledgement of this significant day is another example of the leadership principle "know your Marines."

• *Unit defining moment.* Units should pick other events when they were stressed and tested as an entire unit. Following this challenge, while the unit still shares a sense of accomplishment, is an opportune time for a commander to praise his Marines.

• *Personal correspondence.* Commanders should pen a short note to Marines to recognize significant accomplishments or milestones in their Marines' careers. Congratulations or expressions of sympathy are particularly important.

• *Unit symbol, mascot, logo, or motto.* Many units use a unit symbol, mascot, logo, or motto to give Marines a sense of belonging. These symbols make Marines feel like they are part of a larger and unique entity. This enables them to identify with the unit and aids in developing both horizontal and vertical cohesion.

Unit Training and Exercises. Unit training and exercises are often the best ways to develop horizontal cohesion. The interaction, close living quarters, and shared hardship of these exercises often bring Marines and units together. Once again, unit training and exercises should be either initiated or followed up by a command letter home and a hometown news release.

THE MARINE CORPS FAMILY

The Marine Corps is a family. The intense bond we share with each other makes us all brothers and sisters. These bonds can often become as strong between Marines as blood siblings. We sometimes look to our Marine Corps mentors much like a parent and we never want to let our fellow Marines down. Our heritage is replete with examples of Marines who sacrificed themselves so that they could save the lives of their fellow Marines. No matter whether a Marine has a loving family back home or no family at all, he will always have the support and bond of the Marine Corps family.

Since we are a larger Marine Corps family, we must keep each Marine's family—however it is defined—informed, involved, and supported. Informed Marine families have greater comfort when their Marines deploy, which allows the Marine to concentrate on the mission because his family understands what his job requires. An involved family is an educated family that has a clear understanding of what the Marine Corps asks of them and their loved one in the Marine Corps. A supported family knows that there are Marines and professionals to whom they can turn when they need help.

Family readiness has been an important aspect of sustaining the transformation for some time and has grown significantly in the last decade. Marine leaders need to have an intimate understanding of family readiness and the programs available to units, families, and married and single Marines. Get your Marines and their families involved and be innovative in ways that support the Marine Corps family.

Chapter 7

Phase V: Citizenship

The fifth phase of transformation is citizenship. Beyond preparing a Marine Corps that will win in combat, what truly distinguishes our legacy to our nation are the citizens we produce—citizens transformed by their Marine Corps experience and enriched by their internalization of our ethos, ideals, and values. As Marines, they have learned a nobler way of life, they are able to draw from their experiences, and they are prepared to be leaders within the Corps and within their communities. Our nation's most tangible benefit comes to fruition during this phase of transformation. We produce citizens with our core values—the highest ideals in the American character—and place them in an environment where they are held accountable for those values.

As Lieutenant General Victor H. Krulak wrote in his book, *First to Fight*, Marines "are masters of a form of unfailing alchemy which converts unoriented youths into proud, self-reliant stable citizens—citizens into whose hands the nation's affairs may safely be entrusted." Although our Corps has its share of heroic figures, in the minds of the American people our fame is collective, not individual. Ask the average American to name a famous

Soldier or Sailor, and he will quickly respond with such names as John Paul Jones, Douglas MacArthur, or George Patton. Ask him to name a famous Marine, and they will most likely draw a blank. Yet, to them, the word "Marine" is synonymous with honor, courage, and commitment: our core values. They expect them to rise above self-interest, and they expect them to lead. Their expectations of veteran Marines are the same as those they place on active and reserve Marines. When we "make Marines," we make Marines for life. We provide our Nation with a legacy of productive citizens, transformed by their experiences while on active duty and enriched by their internalization of our ethos, ideals, and values.

Nearly 70 percent of all active duty Marines are first-term enlistees who leave active duty at completion of their first term. While a few will remain and provide our critical NCO and staff NCO leadership, most have other aspirations—yet unfulfilled dreams—and they will depart upon completion of 4 years of faithful service. Though thousands of Marines leave the Corps each year, they will always be United States Marines. They earned that title at the end of recruit training and have been expected to live up to those responsibilities ever since.

The responsibility of being a Marine does not end upon leaving active ranks. In many respects, it only just begins. While these men and women are no longer under the watchful eye of their Marine superiors and no longer subject to the Uniform Code of Military Justice, they continue to be judged by fellow Americans on their actions and the quality of their character for the rest of their lives. This judgment extends not only to a Marine's actions, but also to how he or she represents themselves online. Posts, blogs, and Web pages offer everyone an opportunity to express an opinion, but Marines should consider how such opinions may appear to others and whether those statements are appropriate.

When Marines depart the Corps, they will be in the ideal position to demonstrate that Marines reflect the values that Americans cherish most and hold in the highest regard—the values upon which this nation was founded and which guide us as we shoulder the responsibility of a world power. Be it a 4-year enlistment or a 35-year career, we all must sooner or later take off the uniform, but we have every reason to take great pride in our service. We have done something that few Americans today ever consider doing—we have sacrificed our personal comfort and liberties for the health and needs of the Nation. In return, we were imbued with time-tested values of honor, courage, and commitment that provide the foundation for personal success in any endeavor. These values serve as a moral compass as we return to further education or to join the workforce, and these values will make us leaders in our universities, workplaces, and communities.

Marines leaving active duty are our ambassadors and advocates woven throughout the fabric of America. If the Marine Corps has done its job and the Marine has embraced it completely, the transformation of a Marine will remain central in defining his or her life. The success of the full transformation experience is evident through such experiences as reunions, various Marine Corps fraternal organizations, and the heartfelt and mutually understood *semper fidelis*. We all are tremendously proud: we adorn our vehicles with Marine Corps stickers; wear the eagle, globe and anchor; pause wherever we are on 10 November to celebrate and remember; and jump with exuberant willingness at the chance to share our sea stories and history with others. Eventually, such enthusiasm reaches the eyes and ears of a young person thinking about becoming a Marine, making the cycle of transformation both complete and starting anew.

A positive transformation cycle will help us live up to the responsibility of returning better citizens to the Nation. By instilling the values that we have always held true, we develop Marines who are capable of being solid and contributing citizens.

Chapter 8

Critical Factors Affecting Sustainment

Once Marines arrive at their units, they begin a critical period wherein several opportunities exist for reinforcement or disillusionment. These opportunities will influence their decisions at a number of decision points, such as their impressions of the check-in process, their squad leader, the MOS school, billeting, their NCO in charge, and the chain of command. They may ask themselves, "Do I want to be part of this organization?" Their response will be one of two paths: they will either become "self-sustaining," headed toward successful enlistments and aiding others in sustainment, or they will become "at-risk," headed for first-term, nonexpiration of active service attrition.

OBSTACLE REDUCTION

The commander is responsible for creating a climate that allows junior leaders to effectively apply Marine Corps leadership principles. As Marines enter their unit for the first time, they closely observe the unit and they quickly assess the actions and practices of their leaders and their peers. Experience has taught us that if a Marine feels there are too many obstacles to allow him to integrate into the unit, he will fail. We must reduce these obstacles and foster *esprit de corps*.

Marines take pride in being able to meet challenges and achieve success. Success is founded on a leader's ability to identify an obstacle, recognize and accept the constraints, and find a workable solution. Accepting the constraints is a key factor. All units encounter the reality of operational tempo and manpower shortfalls; limits of time, money, manpower, and other external influences will always exist. Accepting these realities as constraints, instead of obstacles, enables a leader to focus on the actual obstacle at hand. Once obstacles are properly identified, a command can begin to identify means to overcome them.

STRONG LEADERSHIP

Marine Corps history is replete with examples of Marines who have overcome great odds to achieve success. These success stories have a common feature: initiative and ingenuity exercised by leaders at all levels. Our ability to overcome obstacles to sustaining the transformation depends on leaders who can use leadership principles to reduce obstacles.

Our junior leaders determine the obstacles on which a command focuses. Our small-unit leaders conduct the tasks, complete the missions, and ensure unit effectiveness for the command. Their influence and impact on their subordinates' growth as future leaders determine a unit's success in sustaining the transformation. Our small-unit leaders, the "strategic corporals," should be the Corps' primary focus to ensure success both on and off the battlefield. To develop and employ effective junior leaders and small-unit leaders, our efforts must go beyond training to attain tactical and technical proficiency. As leaders, we must also focus on imparting leadership traits—often remembered as JJ DID TIE BUCKLE—to our Marines:

- *Justice*. Giving reward and punishment according to merits of the case in question. It is also the ability to administer a system of rewards and punishments impartially and consistently.

- *Judgment*. The ability to weigh facts and possible solutions on which to base sound decisions.

- *Dependability*. The certainty of proper performance of duty.

- *Initiative*. Taking action in the absence of orders.

- *Decisiveness*. Ability to make decisions promptly and to announce them in a clear, forceful manner.

- *Tact*. The ability to deal with others without creating offense.

- *Integrity*. Uprightness of character and soundness of moral principles. Integrity includes the qualities of truthfulness and honesty.

- *Enthusiasm.* The display of sincere interest and exuberance in the performance of duty.

- *Bearing.* Creating a favorable impression in carriage, appearance, and personal conduct at all times.

- *Unselfishness.* Avoidance of providing for one's own comfort and personal advancement at the expense of others.

- *Courage.* The mental quality that recognizes fear of danger or criticism, but enables a Marine to proceed in the face of it with calmness and firmness.

- *Knowledge.* The range of one's information, including professional knowledge and an understanding of your Marines.

- *Loyalty.* The quality of faithfulness to a Marine's Country, Corps, unit, seniors, subordinates, and peers.

- *Endurance.* The mental and physical stamina measured by the ability to withstand pain, fatigue, stress, and hardship.

With leadership as a focus, we can begin to examine how we can better prepare our young Marines to overcome obstacles. As discussed in the following subparagraphs, to sustain the transformation and create effective units, we must rely on a command's ability to emphasize and reinforce the tried and true principles of leadership.

Be Technically and Tactically Proficient

A technically and tactically proficient Marine knows his job thoroughly and possesses a wide field of knowledge. Before you can lead, you must be able to do the job. Tactical and technical competence can be learned from books and from on-the-job training. To develop this leadership principle of being technically and tactically proficient—

- Know what is expected of you then expend time and energy on becoming proficient at those things.
- Form an attitude early on of seeking to learn more than is necessary.
- Observe and study the actions of capable leaders.
- Spend time with those people who are recognized as technically and tactically proficient at those things.
- Prepare yourself for the job of the leader at the next higher rank.
- Seek feedback from superiors, peers, and subordinates.

Know Yourself and Seek Self-Improvement

This principle of leadership should be developed by the use of leadership traits. Evaluate yourself by using the leadership traits and determine your strengths and weaknesses. You can improve yourself in many ways. To develop the techniques of this principle—

- Make an honest evaluation of yourself to determine your strong and weak personal qualities.
- Seek the honest opinions of your friends or superiors.
- Learn by studying the causes for the success and failures of others.
- Develop a genuine interest in people.

- Master the art of effective writing and speech.
- Have a definite plan to achieve your goals.

Know Your Marines and Look Out for Their Welfare

This is one of the most important of the leadership principles. A leader must make a conscientious effort to observe his Marines and how they react to different situations. A Marine who is nervous and lacks self-confidence should never be put in a situation that requires an important decision. This knowledge will enable you as the leader to determine when close supervision is required. To put this principle in to practice successfully you should—

- Put your Marines welfare before your own.
- Be approachable.
- Encourage individual development.
- Know your unit's mental attitude; keep in touch with what unit members think.
- Ensure fair and equal distribution of rewards.
- Provide sufficient recreational time and insist on participation.

Keep Your Marines Informed

Marines by nature are inquisitive. To promote efficiency and morale, a leader should inform the Marines in his unit of all happenings and give reasons why things are to be done. This is accomplished only if time and security permits. Informing your Marines of the situation makes them feel that they are a part of the team and not just a cog in a wheel. Informed Marines perform better. The key to giving out information is to be sure that the Marines have enough information to do their job intelligently and to inspire their initiative, enthusiasm, loyalty, and convictions. Techniques to apply this principle include the following:

• Explain why tasks must be done and the plan to accomplish a task whenever possible.

• Be alert to detect the spread of rumors. Stop rumors by replacing them with the truth.

• Build morale and *esprit de corps* by publicizing information concerning successes of your unit.

• Keep your unit informed about current legislation and regulations affecting their pay, promotion, privileges, and other benefits.

Set the Example

A leader who shows professional competence, courage, and integrity sets high personal standards for himself before he can rightfully demand it from others. Your appearance, attitude, physical fitness, and personal example are all on display daily for the Marines and Sailors in your unit. Remember, your Marines and Sailors reflect your image! Techniques for setting the example are to—

• Show your subordinates that you are willing to do the same things you ask them to do.

• Maintain an optimistic outlook.

- Conduct yourself so that your personal habits are not open to criticism.
- Avoid showing favoritism to any subordinate.
- Delegate authority and avoid over supervision in order to develop leadership among subordinates.
- Teach leadership by example.

Ensure Assigned Tasks are Understood, Supervised, and Accomplished

Leaders must give clear, concise orders that cannot be misunderstood, and then closely supervise to ensure that these orders are properly executed. Before you can expect your Marines to perform, they must know what is expected of them. The most important part of this principle is the accomplishment of the mission. In order to develop this principle—

- Issue every order as if it were your own.
- Use the established chain of command.
- Encourage subordinates to ask questions concerning any point in your orders or directives they do not understand.
- Question subordinates to determine if there is any doubt or misunderstanding in regard to the task to be accomplished.
- Supervise the execution of your orders.
- Exercise care and thought in supervision; over supervision will hurt initiative and create resentment, while under supervision will not get the job done.

Train Your Marines as a Team

Teamwork is the key to successful operations and is essential from the smallest unit to the entire Marine Corps. As a leader, you must insist on teamwork from your Marines. Train, play, and operate as a team. Be sure that each Marine knows his position and responsibilities within the team framework. To develop the techniques of this principle—

- Stay sharp by continuously studying and training.
- Encourage unit participation in recreational and military events.
- Do not publicly blame an individual for the team's failure or praise just an individual for the team's success.
- Ensure that training is meaningful, and that the purpose is clear to all members of the command.
- Train your team based on realistic conditions.
- Insist that every person understands the functions of the other members of the team and the function of the team as part of the unit.

Make Sound and Timely Decisions

The leader must be able to rapidly estimate a situation and make a sound decision based on that estimation. Hesitation or a reluctance to make a decision leads subordinates to lose confidence in your abilities as a leader. Loss of confidence in turn creates confusion and hesitation within the unit. Techniques to develop this principle include the following:

- Develop a logical and orderly thought process by practicing objective estimates of the situation.

- Plan for every possible event that can reasonably be foreseen when time and situation permit.

- Consider the advice and suggestions of your subordinates before making decisions.

- Consider the effects of your decisions on all members of your unit.

Develop a Sense of Responsibility Among Your Subordinates

Another way to show your Marines you are interested in their welfare is to give them the opportunity for professional development. Assigning tasks and delegating authority promotes mutual confidence and respect between leader and subordinates. It also encourages subordinates to exercise initiative and to give wholehearted cooperation in accomplishment of unit tasks. When you properly delegate authority, you demonstrate faith in your Marines and increase authority, and you increase their desire for greater responsibilities. To develop this principle—

- Operate through the chain of command.

- Provide clear, well thought out directions.

- Give your subordinates frequent opportunities to perform duties normally performed by senior personnel.

- Be quick to recognize your subordinates' accomplishments when they demonstrate initiative and resourcefulness.
- Correct errors in judgment and initiative in a way that will encourage the individual to try harder.
- Give advice and assistance freely when your subordinates request it.
- Resist the urge to micromanage.
- Be prompt and fair in backing subordinates.
- Accept responsibility willingly and insist that your subordinates live by the same standard.

Employ Your Command in Accordance with its Capabilities

A leader must have a thorough knowledge of the tactical and technical capabilities of the command. Successful completion of a task depends upon how well you know your unit's capabilities. If the task assigned is one that your unit has not been trained to do, failure is very likely to occur. Failures lower your unit's morale and self-esteem. Seek out challenging tasks for your unit, but be sure that your unit is prepared for and has the ability to successfully complete the mission. Techniques for development of this principle include the following:

- Avoid volunteering your unit for tasks that are beyond their capabilities.
- Be sure that tasks assigned to subordinates are reasonable.
- Assign tasks equally among your subordinates.
- Use the full capabilities of your unit before requesting assistance.

Seek Responsibility and Take Responsibility for Your Actions

For professional development, you must actively seek out challenging assignments. You must use initiative and sound judgment when trying to accomplish jobs that are required by your grade. Seeking responsibilities also means that you take responsibility for your actions. Regardless of the actions of your subordinates, the responsibility for decisions and their application falls on you. Techniques in developing this principle include the following:

- Learn the duties of your immediate senior, and be prepared to accept the responsibilities of these duties.

- Seek a variety of leadership positions that will give you experience in accepting responsibility in different fields.

- Take every opportunity that offers increased responsibility.

- Perform every task, no matter whether it is classified or seemingly trivial, to the best of your ability.

- Stand up for what you think is right. Have courage in your convictions.

- Carefully evaluate a subordinate's failure before taking action against that subordinate.

- In the absence of orders, take the initiative to perform the actions you believe your senior would direct you to perform if present.

Everyone in the Corps either knows or remembers some of these principles, but many do not consistently and universally apply them. Obstacles to sustainment are obstacles to the application of our leadership principles.

Interacting with Schools

The benefits of interaction between commands and schools cannot be understated. Any effort commands can make toward easing the transition from schools to units will better prepare Marines to meet upcoming challenges.

Attending Unit-Level Corporal Courses

Upon promotion to corporal, some Marines understand their new grade and responsibilities by their observations of the corporals with whom they have served. Commands have an opportunity and responsibility to develop NCOs as leaders to ensure they are prepared to take on leadership challenges. Many units have implemented a 2-day corporal's course conducted by the command's staff NCOs. Though the course can be tailored to the unit's requirements, guidance in such areas as professional relations with subordinates, the NCO's role in sustaining the transformation, counseling, core values, mentoring, and applying leadership principles provide a sound starting point. This course must emphasize to newly promoted corporals that promotion to their new grade not only places them in charge of those in their unit, but also in charge of all Marines of lesser grade.

Encouraging Professional Military Education

Enlisted professional military education is another tool provided by the Corps to develop Marine leaders. While not mandatory for promotion, the long-term benefits of formal education for Marines and their unit outweigh the short-term loss of the individual at the unit.

Briefing New Joins

A new join briefing is an opportunity for the command to welcome aboard its new Marines and to provide some insight into the unit itself. Understanding the origin of unit traditions and history can initiate a sense of belonging and unit pride. This briefing should also ensure that Marines have a clear understanding of their new chain of command and how subordinate units fit within senior units.

Maintaining Bachelor Enlisted Quarters

The Commandant's bachelor enlisted quarters campaign plan is dedicated to providing Marines with living conditions that allow them to continue to develop as Marines. Marines' quarters should also foster unit integrity. Every quarters of our Marines are expected to be supervised with engaged leadership. There is clear policy outlining what is expected, but the end result must be quarters where every Marine feels safe. Supportive and engaged leadership also ensures reasonable health and comfort standards are being met.

Providing Mentors

A Marine's attitudes, ethics, and traits frequently conform to those displayed by the role models provided by their unit. Commanders should ensure that the unit provides positive, quality mentors. Mentorship is voluntary but every Marine should look for opportunities to help another Marine attain his future goals, both in or out of the Corps.

Educating Leaders

Recent changes to recruit training and the SOI have led to many misconceptions, one being that training is not as good or as tough as it used to be. These misconceptions cause negative attitudes and lowered expectations. To dispel these misconceptions, former drill instructors and recruiters in the command can provide current and accurate information. Accurate information enables Marines to recognize that while the training has changed and new terms are used, our entry-level training has the same focus—we still make Marines.

This Page Intentionally Left Blank

Glossary

Acronyms

MOS . . . . military occupational specialty

NCO . . . . noncommissioned officer

PFC . . . . private first class

SOI . . . . School of Infantry

US . . . . United States

Terms and Definitions

cohesion—The intense bond developed among Marines in a unit, resulting in absolute trust, subordination of self, understanding of the unit's collective actions, and the importance of teamwork, resulting in increased combat power. (It is the third phase of transformation.) (This term and definition are for the purpose of this Marine Corps reference publication only.)

horizontal cohesion—The horizontal bonding (also known as peer bonding) that builds on a sense of trust and familiarity among individuals of the same rank or position. (This term and definition are for the purpose of this Marine Corps reference publication only.)

sustainment—Sustainment is the responsibility of unit leaders to maintain and build upon the values and warrior spirit built by formal schools and entry-level training. The fourth and most critical phase of both the transformation and the Marine Corps Values Program. (This term and definition are for the purpose of this Marine Corps reference publication only.)

transformation—An ongoing process that begins with prospective enlisted's first contact with a Marine recruiter and continues throughout the Marine's entire life. Transformation has five phases: recruiting, recruit training, cohesion, sustainment, and citizenship. (This term and definition are for the purpose of this Marine Corps reference publication only.)

vertical cohesion—A dimension of cohesion between subordinate and senior that develops from mutual trust and unity. (This term and definition are for the purpose of this Marine Corps reference publication only.)

References

Robert Debs Heinl, Jr., Colonel, USMC, Retired, *Dictionary of Military and Naval Quotations* (Annapolis, MD: United States Naval Institute, 1966).

Krulak, General Victor H. *First to Fight: An Inside View of the U.S. Marine Corps* (Annapolis, MD: Naval Institute Press, 1984).

CPSIA information can be obtained
at www.ICGtesting.com
Printed in the USA
LVHW051553151121
703392LV00012B/1605

9 781540 870988